this book belongs to

a Paul Doodles original

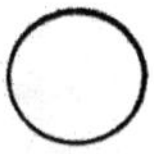

Time to get a new notebook

www.ingramcontent.com/pod-product-compliance
Lightning Source LLC
Chambersburg PA
CBHW081602220526
45468CB00010B/2742
* 9 7 8 1 7 9 8 7 1 8 3 8 4 *